UNNATURAL DISASTERS

WASTED WATERS

By Emily Mahoney

Gareth Stevens
PUBLISHING

Please visit our website, www.garethstevens.com. For a free color catalog of all our high-quality books, call toll free 1-800-542-2595 or fax 1-877-542-2596.

Cataloging-in-Publication Data
Names: Mahoney, Emily.
Title: Wasted waters / Emily Mahoney.
Description: New York : Gareth Stevens Publishing, 2018. | Series: Unnatural disasters | Includes index.
Identifiers: ISBN 9781538205273 (pbk.) | ISBN 9781538205297 (library bound) | ISBN 9781538205280 (6 pack)
Subjects: LCSH: Water–Pollution–Juvenile literature. | Water conservation–Juvenile literature.
Classification: LCC TD422.M34 2018 | DDC 363.739'4–dc23

First Edition

Published in 2018 by
Gareth Stevens Publishing
111 East 14th Street, Suite 349
New York, NY 10003

Designer: Sam DeMartin
Editor: Joan Stoltman

Photo credits: Cover, p. 1 Ed Wray/Getty Images News/Getty Images; p. 5 Triff/Shutterstock.com; pp. 7, 29 Pro_Vector/Shutterstock.com; p. 9 (photo) PLANETOBSERVER/Getty Images; p. 9 (map) Rainer Lesniewski/Shutterstock.com; p. 10 U.S. Navy/Getty Images News/Getty Images; p. 11 Benjamin Lowy/Getty Images News/Getty Images; p. 13 Bettmann/Bettmann/Getty Images; p. 14 David McNew/Getty Images News/Getty Images; p. 15 David McNew/Getty Images News/Getty Images; p. 17 (top) Newspix/Newspix/Getty Images; p. 17 (bottom) Irfan Khan/Los Angeles Times/Getty Images; p. 18 © iStockphoto.com/PhilAugustavo; p. 19 Michael Pitts/Nature Picture Library/Getty Images; p. 21 Barcroft/Barcroft Media/Getty Images; p. 23 (all) courtesy of NASA; p. 24 Mark Wilson/Getty Images News/Getty Images; p. 25 Bill Pugliano/Getty Images News/Getty Images; p. 26 JIM WATSON/AFP/Getty Images; p. 27 Brett Carlsen/Getty Images News/Getty Images.

Printed in the United States of America

CPSIA compliance information: Batch #CS17GS: For further information contact Gareth Stevens, New York, New York at 1-800-542-2595.

CONTENTS

Words in the glossary appear in **bold** type the first time they are used in the text.

OUR DAMAGED PLANET

Earth has been increasingly ravaged, or damaged, by industrial accidents and mistreatment. Some might say people are the deadliest creatures on Earth because of the damage we've caused to the land and water that make up our planet. While nobody intends to destroy nature, people have had a very negative effect on Earth and its resources. One of the most evident ways is how we treat Earth's water.

Lakes, rivers, and coastlines all over the world have been polluted. Even our oceans have been poisoned! Water pollution has become an increasingly problematic issue that affects more and more people and animals. Read on to learn about specific incidents of wasted waters, as well as what you can do to help stop this terrible problem from getting worse.

A WATERY WORLD

- About 70 percent of Earth is covered in water!
- The oceans hold about 97 percent of Earth's water.
- Less than 4 percent of Earth's water is freshwater!
- Globally, 70 percent of water sources are used for agriculture and irrigation.
- Only 10 percent of water is used domestically, or in homes.
- As many as one in every nine people worldwide don't have **access** to safe and clean drinking water!

LASTING EFFECT

The Environmental Protection Agency, or EPA, is a government agency in the United States. Its job is to protect water, air, and land by helping make laws, inspecting how factories throw out or store waste, taking polluters to court, and more.

Since water is necessary for people, animals, and plants to live, it's very important that the whole world works together to keep water pollution-free!

WHAT'S IN OUR WATER?

Each molecule of water is chemically made up of two parts hydrogen and one part oxygen. A single drop of water contains billions of water molecules. Naturally, some other matter enters our water, such as minerals, soil, and even tiny parts of rocks or leaves, but these things are filtered out before water reaches our sinks and bathtub at home. But how do dangerous chemicals or minerals enter our water?

Matter such as oil, **heavy metals**, and other harmful substances sometimes get into our water supply. Sometimes, an accident spills toxins into a water supply. Other times, toxins slowly drain through the ground little by little over many years, as when chemicals are sprayed onto crops to keep pests away.

LASTING EFFECT

When **groundwater** is contaminated, or polluted, it spreads throughout the local water supply, poisoning people, plants, and animals. Groundwater spreads toxins further than a disaster site, ruining an area for decades. Plus, once groundwater is affected, it's often impossible to fix.

When a toxin is found in a water supply, it's measured in tests by ppb, or the number of toxic parts per billion water parts. So, 15 ppb of lead means that 15 molecules of lead were found for every 1 billion water molecules.

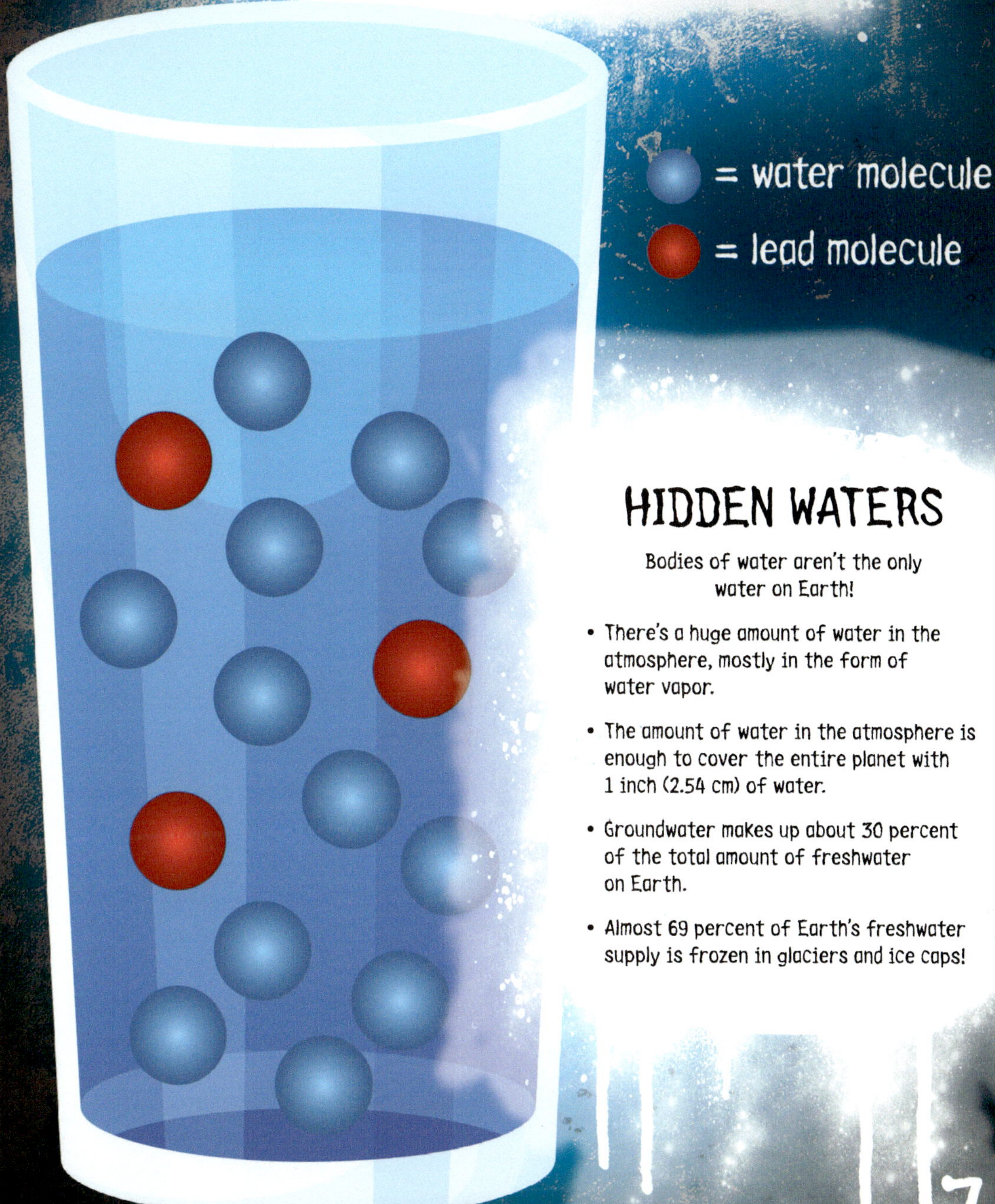

HIDDEN WATERS

Bodies of water aren't the only water on Earth!

- There's a huge amount of water in the atmosphere, mostly in the form of water vapor.
- The amount of water in the atmosphere is enough to cover the entire planet with 1 inch (2.54 cm) of water.
- Groundwater makes up about 30 percent of the total amount of freshwater on Earth.
- Almost 69 percent of Earth's freshwater supply is frozen in glaciers and ice caps!

ALGAE CAN KILL

The Gulf of Mexico is a popular spot for beaches, boating, and tourism. However, it's also home to a serious pollution problem! Every year, heavy rains and melting snows wash large amounts of nitrogen and phosphorus from lawns, sewage treatment plants, and farmland into the Gulf of Mexico by way of the Mississippi River. The river runs for around 2,300 miles (3,700 km) through the United States and drains into the gulf.

LASTING EFFECT

The size of the dead zone in the Gulf of Mexico changes every year and is expected to become larger if something isn't done!

Nitrogen and phosphorus are **nutrients** used to help lawns and crops grow. However, when they enter the gulf, they cause large amounts of algae growth, or algae blooms. Algae blooms take all the oxygen from the water, making parts of the gulf difficult, if not impossible, for marine life to survive in. These areas are called dead zones!

DEAD ZONE FACTS

- The marine life in the Gulf of Mexico is a major source of money and jobs, supplying about 70 percent of all shrimp and 65 percent of all oysters in the United States.
- About 40 percent of US land (not including Alaska and Hawaii) drains into the Mississippi River!
- The Gulf of Mexico dead zone was over 6,400 square miles (16,600 sq km) in 2015, over twice the size of Delaware!

The Gulf of Mexico is surrounded by the United States, Mexico, and Cuba, but little is reported about the effects the dead zone is having on Mexico or Cuba!

THE GLOOMY GULF

The dead zone isn't the only man-made disaster that's affected the **environment** of the Gulf of Mexico. Multiple oil spills have also taken place. On April 20, 2010, the largest marine oil spill in history occurred when the Deepwater Horizon oil rig exploded and sank. A huge hole opened in the ocean floor, pouring toxic oil into the gulf for months. Several plans to stop the oil were tried. They all failed.

The Deepwater Horizon oil rig was around 40 miles (64 km) off the coast of Louisiana, but the oil it leaked affected the coastlines of Texas, Mississippi, Alabama, Florida, and six states in Mexico!

THE DESTRUCTION OF DEEPWATER HORIZON

- **people affected:** 11 killed in rig explosion
- **jobs affected:** thousands of fishing industry jobs lost; 8,000 to 12,000 jobs lost when offshore drilling was stopped (though it eventually started again)
- **land affected:** over 1,000 miles (1,600 km) of shoreline polluted
- **water affected:** almost 89,000 square miles (230,500 sq km) of fishing area poisoned and temporarily shut down
- **animals affected:** unknown thousands of fish, sea horses, lobsters, sea stars, oysters, clams, corals, and other sea life killed

Eventually, in September 2010, the oil well was sealed off so oil would stop leaking. But it was too late for much of the area's wildlife. Over 82,000 birds may have been harmed or killed, as well as about 6,000 sea turtles and about 26,000 marine mammals, including dolphins and whales.

LASTING EFFECT

In 2015, oil from the spill was discovered across more than 1,200 square miles (3,100 sq km) of the seafloor. Even today, there hasn't been enough time for many species to recover from the disaster.

A NOT-SO-LOVELY CANAL

Some unnatural disasters affect a huge area, like the Deepwater Horizon oil spill. Others, like the Love Canal disaster, are smaller, but no less harmful. From 1942 to 1953, the Hooker Chemical Company used the unfinished canal near Niagara Falls, New York, as a chemical waste dump—with government permission. The company then sold the land to the Niagara Falls School Board. An elementary school opened in 1955, and soon a neighborhood of hundreds of families grew up around the school.

LASTING EFFECT

In September 1983, the EPA listed Love Canal on its National Priorities List (NPL) and began to work with the state to clean up the site. The cleanup was completed in 2004.

Many residents were upset that they had to leave their home and even more upset that the government took so long to clean up the toxic mess!

In the 1970s, residents noticed they had abnormally high rates of **birth defects**, breathing problems, severe headaches, and other unexplained illnesses. It was discovered the toxic chemicals dumped into the canal were in yards, basements, and the groundwater. The government was slow to act, prompting anger from **activists** across the nation.

LOVE CANAL KEY INFORMATION

- **what:** chemical dumping poisons local groundwater
- **poison:** at least 12 chemicals that cause **cancer** and other illnesses
- **how much:** around 21,000 tons (19,000 mt) of toxic chemicals dumped
- **people affected:** around 800 families
- **long-term effects:** around $17 million spent moving families out of the area; 239 families were moved in 1978; the rest weren't moved until 1981

A DESERT DISASTER

Across the country, in the middle of the Mojave Desert, another major water pollution event was poisoning a small American town at the same time as Love Canal's poisoning. A company had dumped roughly 370 million gallons (1.4 billion L) of water polluted with the heavy metal chromium-6 into unlined ponds around the town of Hinkley, California.

Chromium-6 was being used in Hinkley to stop pipes from rusting at the Pacific Gas and Electric power plant, shown here.

LASTING EFFECT

In 2015, the local water board ordered a thorough cleanup of chromium-6 from the water supply, nearly 7 decades after the contamination began and nearly 3 decades since the contamination was first exposed.

Hinkley was a small town with a large problem: Heavy metals like chromium-6 can cause cancer and other illnesses as they build up in a person's body. But people didn't know about the toxic ponds or that the groundwater was increasingly poisonous until 1987. By the time the town discovered what was making so many sick, many people had already been exposed. In fact, they'd been exposed to the poison since the 1950s!

HINKLEY'S NUMBERS

- **what:** groundwater contamination from poor toxic waste disposal
- **poison:** chromium-6, a heavy metal
- **when:** 1950s to present
- **where:** Hinkley, California, the state with the strictest environmental laws in America
- **how much:** as high as 580 ppb
- **how much is normal:** California allowed 50 ppb of chromium-6 at the time
- **long-term effects:** Hinkley went from 8,000 to less than 1,000 residents in 60 years; the town will probably be totally abandoned by the next decade

About 250 acres (100 ha) of alfalfa and grass have been planted in Hinkley to filter toxins out of the local soil.

Hinkley's high number of illnesses caught the attention of a legal clerk named Erin Brockovich when some paperwork came to a lawyer she was working for. She decided to investigate further. In 1993, her law firm sued the company that contaminated the local water supply. When all was said and done, the company was told to pay the victims, stop their use of chromium-6, and clean up the damage they had done in the Hinkley area.

WHAT HAPPENED TO HINKLEY?

The residents of Hinkley were told their town would be cleaned up. But in 2010, they found out that the toxins continued to spread through groundwater. The contamination area is now three times bigger than it was 2 decades earlier when this disaster was detected! The area continues to grow, spreading around 2.5 feet (0.76 m) a day. In 2013, when the last school in Hinkley closed, it became clear the town wouldn't survive this disaster after all.

LASTING EFFECT

Erin Brockovich's lawsuit led to a $333 million payout to 600 victims. After the law firm was paid $133 million, each victim received $333,000.

Brockovich's story gained national attention, and a very successful movie, called *Erin Brockovich*, was released in 2000. The film is credited with drawing attention not just to Hinkley, but to how serious water pollution is. More people became aware of environmental issues and activism because of it.

Erin Brockovich

Brockovich and her team still work in Hinkley to continue the cleanup effort and travel all over the country to help other communities fight for environmental cleanups.

THE GREAT PACIFIC GARBAGE PATCH

Do you know where your trash goes? Some of it may actually end up in the ocean! The Great Pacific Garbage Patch is a huge mess of pollution that floats on top of and in the North Pacific Ocean. No one knows how big it really is! All sorts of trash continues to get stuck in this North Pacific Ocean **gyre** because of the way water currents move. This makes the patch grow bigger and bigger every day!

The area in the center of the gyre is usually very calm, so once trash is caught there, it won't move unless people go out and clean it up! Most of the material in this garbage patch is made up of small bits of plastic, which won't ever break down.

This barrier stops garbage off the coast of Hawaii from becoming a part of the Great Pacific Garbage Patch.

ALL ABOUT THE GARBAGE PATCH

- About 80 percent of the debris in the Great Pacific Garbage Patch comes from people in North America and Asia.
- Trash from North America takes about 6 years to reach the garbage patch, while trash from Asia takes about 1 year.
- The remaining 20 percent of debris in the Great Pacific Garbage Patch comes from boats, offshore oil rigs, and large cargo ships that dump or lose waste directly into the water.

seal caught in the Garbage Patch

LASTING EFFECT

This plastic debris harms many different kinds of animals in the ocean environment. At least 265 different species, or kinds of animals, are affected.

THE WORLD'S MOST POLLUTED RIVER

The Citarum River, located in Indonesia, has been called the most polluted river in the world. It contains massive amounts of trash, or garbage, even though it's the only source of drinking water for millions of people. This means the pollution doesn't just sicken local animals and river life, but also the entire population of people who live near it.

The pollution started because people living near the river didn't have anywhere to get rid of their garbage. So they threw it into the river, not understanding they were only poisoning themselves! Many factories dumped waste into the river, too, as Indonesia has no laws preventing this. In some places, the water can't even be seen because of the amount of trash in the river.

DANGEROUS WATER

- In 2012, scientists reported the wastewater being dumped into the Citarum from one local factory was so toxic it could burn skin!
- Indonesia only regulates 264 chemicals, though over 100,000 are used in factories that dump along the river!
- Close to 5 million people live along the river's 190-mile (305 km) length, drinking, bathing, cooking, and farming with the toxic water!

LASTING EFFECT

In 2010, the Indonesian government began cleaning up the river. It's expected to cost $4 billion and take at least 15 years.

The Citarum River needs more than a cleanup effort. Because people in Indonesia don't know about the effects caused by garbage, or waste, in the river, a big part of the cleanup effort needs to involve educating people about how to deal with trash. Laws need to be developed to stop factory waste from being dumped, too!

THE SHRINKING SEA

Some water problems aren't caused by chemicals. The Aral Sea is located in central Asia, and it was, at one point, the world's fourth-largest saltwater lake. The two rivers that feed into it are the Amu Dar'ya and Syr Dar'ya. In the 1960s, the government decided to change the course of those rivers so nearby desert regions could be used for farming.

But there were consequences for the sea. Less water being brought to the sea meant the amount of salt in the water rose sharply. Also, the water at the top of the sea became much less salty than the water at the bottom. This has meant the sea's surface heats and cools much faster than before, which changes local weather, causing strong winds and dust storms.

A SALTY PROBLEM

- From 1960 to 2000, the average summer monthly air temperature around the Aral Sea increased 7°F (4°C)!
- The course change was done by building a huge irrigation network, with 20,000 miles (32,000 km) of canals, 45 dams, and more than 80 **reservoirs**.
- The eastern part of the Aral Sea dried up completely in 2014 because the temperature change has increased the rate of **evaporation** to an unnaturally fast speed!

Images from space taken just this century alone have captured the visible changes in the Aral Sea.

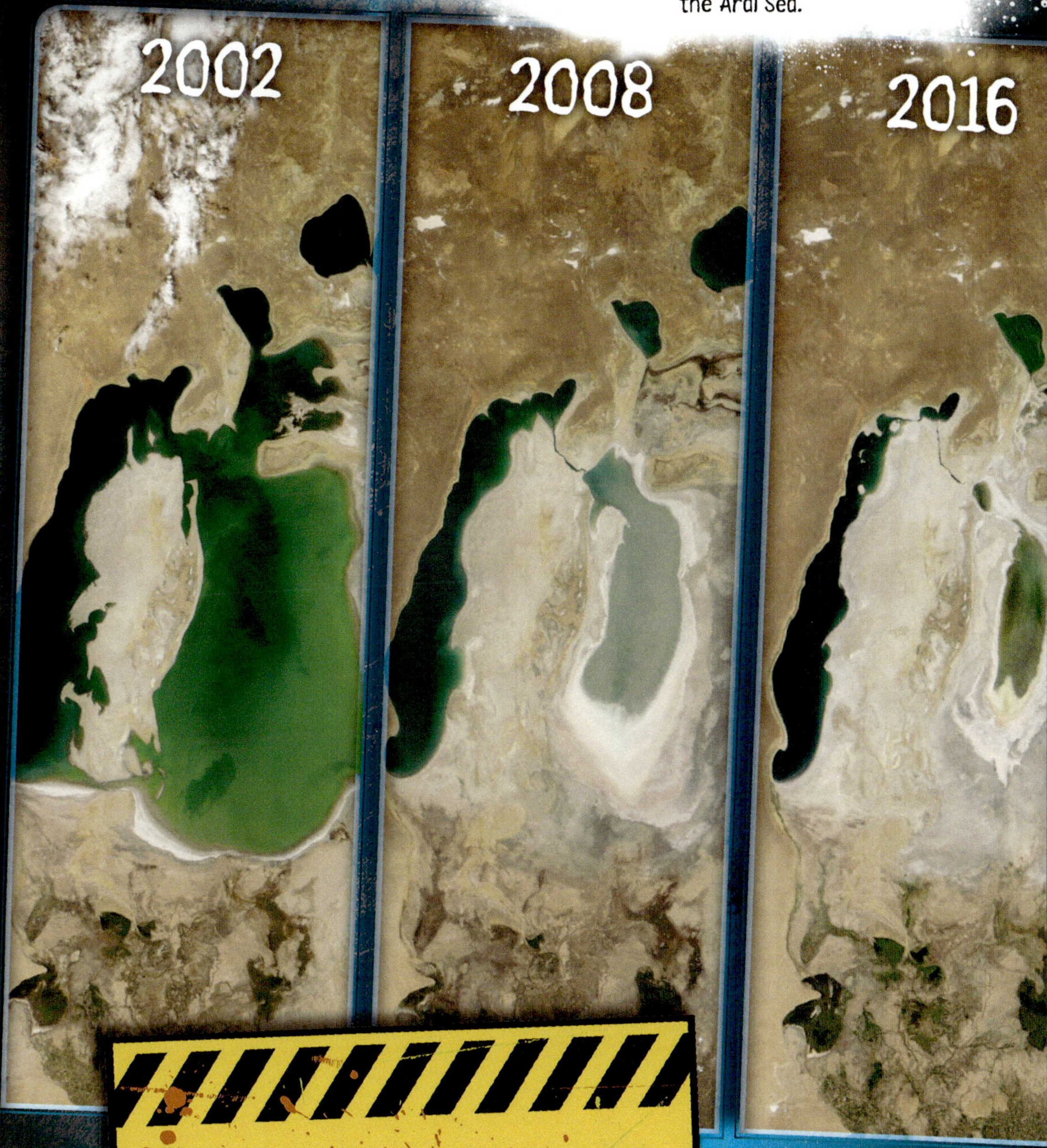

LASTING EFFECT

To help stop the problems of the Aral Sea, a system of dams and canals was built in the early 2000s to put water back into the sea! This has helped raise water levels.

THE FLINT WATER CRISIS

FAST FACTS ON FLINT

- **what happened:** lead and bacteria entered a city water system
- **when:** April 25, 2014, to present
- **how much:** one water test showed 397 ppb of lead (the federal limit is 15 ppb)
- **people affected:** around 98,000 residents, including 8,000 children, poisoned for months
- **short-term effects:** rashes; 12 killed by Legionnaires' disease (caused by bacteria)
- **long-term effects:** serious **neurological** problems likely to develop in children who were age 5 or less when exposed to lead

In 2014, the city of Flint, Michigan, made headlines and shocked the nation with a serious water crisis. In April, Flint switched its water source from Lake Huron to the Flint River, and residents weren't happy. The water was brown, smelly, and tasted weird. Officials told everyone not to worry.

Flint families traveled to Washington, DC, to make sure that Congress saw in person how contaminated Flint's water was.

Months later, residents learned city officials weren't putting phosphates into the water. Phosphates protect metal pipes from breaking down. Without phosphates, the old lead pipes in Flint were eaten away by the water they carried, flaking toxic heavy metal into the water supply in deadly amounts. This toxic water was piped throughout the city to all schools, businesses, and homes. In October 2015, Flint returned to using cleaner Lake Huron water, but the damaged pipes still flake metal.

LASTING EFFECT

Families in Flint now have to rely on bottled water for everything from brushing teeth and bathing to cooking and drinking. This could mean another environmental disaster for Flint if millions of plastic bottles don't get recycled!

Over 30,000 lead pipes need to be replaced, but in 33 months, only 780 have been. It may cost over $55 million and take years to rebuild Flint's water system!

Flint residents were angry, but not just because they were being poisoned. They were upset because it took a long time for anyone to listen to their concerns. One home tested at such high levels of lead that the water could be considered hazardous waste! Residents went to town meetings, wrote to their local and state representatives in Congress, and even wrote to the EPA, but nothing worked.

LASTING EFFECT

The EPA, the state, and the governor of Michigan are all being taken to court. In fact, 13 Flint government officials have been sent to court to decide if they lied. Two of them could go to prison for decades!

The president finally issued a state of emergency in Flint in January 2016, and bottled water was sent from concerned people and celebrities across the country. But for many Flint residents, it was all too little, too late: They had already been exposed.

The children of Flint need yearly blood testing because the poisoned water was pumped into their homes and schools and because lead harms children the worst. Some Flint residents have decided to go to court to get the yearly blood tests for all Flint children paid for.

A LAW THAT SHOULD HAVE STOPPED THIS

The Safe Drinking Water Act is a law that makes states, except Wyoming and Washington, DC, test and write yearly reports about what's in their water. Flint officials lied on their report, so the EPA had to make their own reports. The EPA said there was definitely something wrong, but couldn't help until their report was finished, which took months. With Flint officials lying and the EPA not acting quickly enough, the poisoning continued for 18 months straight before anything was done.

A CALL TO ACTION

WE STILL NEED HELP!

- At least 320 million people in China don't have access to clean drinking water!
- Around 250 million people worldwide are killed yearly by diseases caused by water pollution!
- In America, 40 percent of all rivers and 46 percent of all lakes are polluted and considered unhealthy for swimming, fishing, or aquatic life!
- Around 70 percent of the world's industrial waste is dumped into bodies of water used by people for drinking, bathing, and more!

Much can be done to prevent future problems with one of Earth's most precious resources, water. Many people live with unsafe water every day, but taking action now can help make certain future generations won't have to endure these same kinds of water disasters.

Filtering has become a great option for cleaning dirty water and has improved over the years because of new inventions. But solutions must also involve stopping pollution, not just cleaning up what has already been done. Laws must not only be written, but also enforced and followed! If we work together to commit to preserving this important natural resource, water pollution can be a thing of the past—and should be! No person should have to fear if the water coming into their home is safe.

HOW CAN I HELP EARTH'S

WASTED WATERS?

There are several things you can do both in your home and at school to help keep your local water clean.

- Write your local and state government officials, members of Congress, or your state's governor, requesting that they support antipollution bills and laws!
- Be careful about what cleaning supplies go down your drains and toilets!
- Dispose of trash properly!
- Recycle, help others recycle, or start a recycling program at your school!
- Educate your family and friends about the use of chemicals, such as cleaning supplies and lawn and gardening products!
- Work with a group to clean up trash in your local water!

LASTING EFFECT

Water is necessary for life. If we keep polluting it, people, animals, and plants will continue to be sickened and killed.

GLOSSARY

access: a way of being able to use or get something

activist: one who acts strongly in support of or against an issue

birth defect: a physical or chemical problem that exists at birth and is either passed down from the parents or caused by exposure to a toxin

cancer: a disease caused by the uncontrolled growth of cells in the body

environment: the conditions that surround a living thing and affect the way it lives

evaporation: the change of a liquid to a gas

groundwater: water held underground in the soil or in small holes in rock

gyre: a giant, rotating ocean current

heavy metal: a naturally occurring element that is very toxic to humans and animals

neurological: having to do with the nervous system, including the brain

nutrient: something a living thing needs to grow and stay alive

reservoir: a place where something is stored

FOR MORE INFORMATION

BOOKS

Brennan, Linda Crotta. *The Gulf Oil Spill.* Minneapolis, MN: ABDO Publishing Company, 2014.

Levete, Sarah. *Destroying the Oceans.* New York, NY: Crabtree Publishing Company, 2010.

Newman, Patricia. *Plastic Ahoy! Investigating the Great Pacific Garbage Patch.* Minneapolis, MN: Millbrook Press, 2014.

Websites

Water Pollution Guide

water-pollution.org.uk/

Learn more about water pollution by checking out these helpful diagrams!

What Is Your Water Footprint?

environment.nationalgeographic.com/environment/freshwater/change-the-course/water-footprint-calculator/

Participate in this online tour to see how much water you use and how you can use less!

Publisher's note to educators and parents: Our editors have carefully reviewed these websites to ensure that they are suitable for students. Many websites change frequently, however, and we cannot guarantee that a site's future contents will continue to meet our high standards of quality and educational value. Be advised that students should be closely supervised whenever they access the Internet.

INDEX